Animal Engineers

Izzi Howell

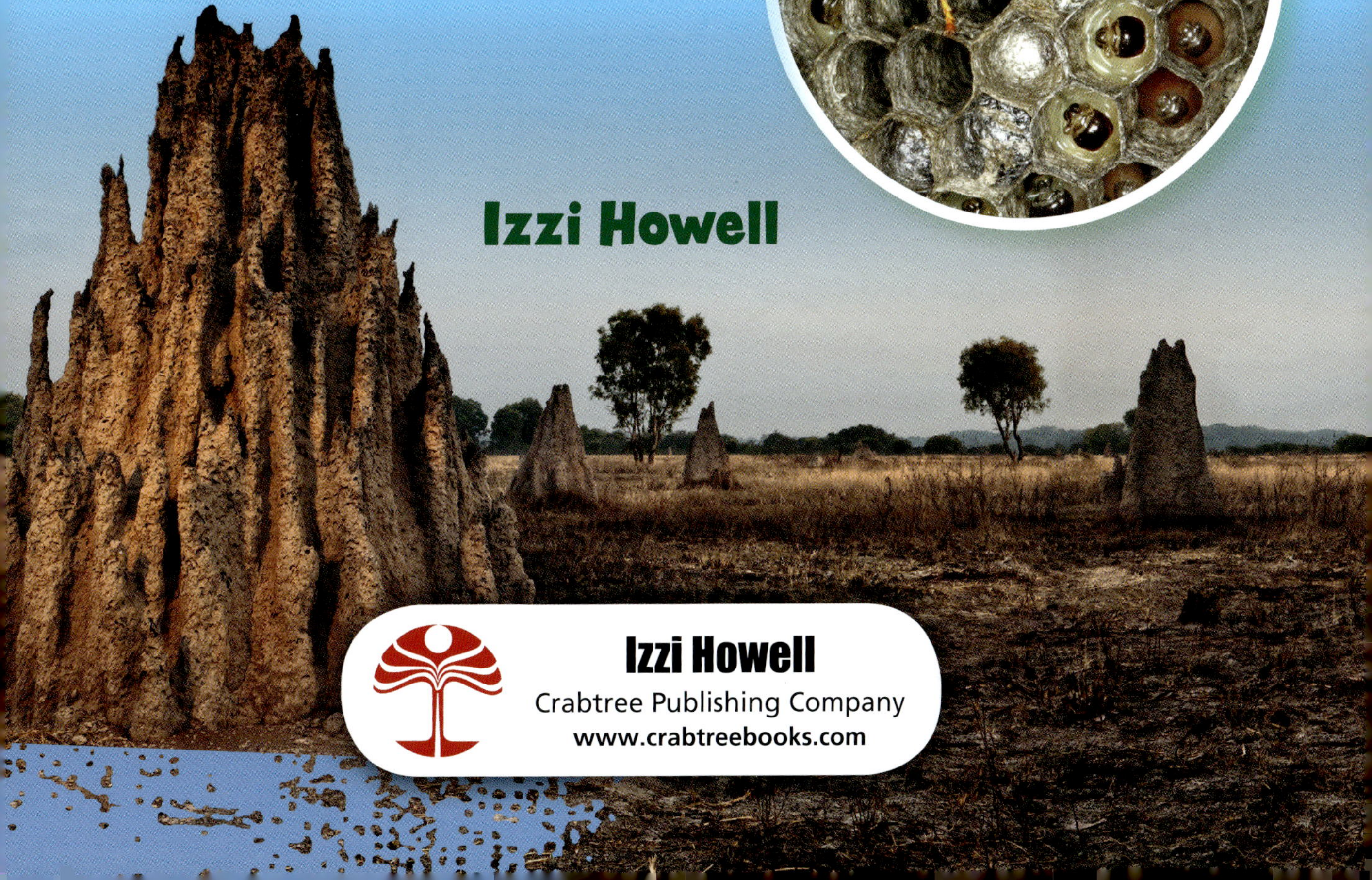

Izzi Howell
Crabtree Publishing Company
www.crabtreebooks.com

CRABTREE
PUBLISHING COMPANY
WWW.CRABTREEBOOKS.COM

Author:
Izzi Howell
Editorial director:
Kathy Middleton
Editor:
Izzi Howell
Proofreaders:
Sonya Newland, Crystal Sikkens
Graphic design:
Richard Parker
Image research:
Izzi Howell
Production coordinator and prepress:
Tammy McGarr
Print coordinator:
Katherine Berti

Images:
Alamy: 8 (Minden Pictures), 19b (Avalon/Photoshot License);
Getty Images: title page top and 21b (MyImages_Micha), 3 and 28t (cinoby), 4 (PCHT), 5 (stanley45), 6 (praisaeng), 10 (MikeLane45), 11tl (Mike Powles), 13t (grandaded), 17b (Besjunior), 19t (Mastamak), 21c (benedek), 22 (Fred Bruemmer), 23t (ugurhan), 24 (David Butler), 25b (wrangel), 26t (pierivb), 27t (Atypeek), 27cl (DON EMMERT/AFP), 27cr (arousa), 28b (UWPhotog), 29t (vlad61), 29b (crispyfrog); Nature Picture Library: 18 (Barrie Britton); Shutterstock: title page centre and 13c (Shishka4), title page bottom and 14 (Andrew Paul Deer), 7l (Independent birds), 7tr (kajornyot wildlife photography), 7br (yusuf madi), 9t (Abeselom Zerit), 9b (IfSea), 11tr (Mark Bridger), 11bl (COULANGES), 12 (Le Do), 13b (OUESTUSA), 15tr (crbellette), 15l (khlungcenter), 15br (S-F), 16 (Steven Ellingson), 17tl (Luc Pouliot), 17tr (IanRedding), 17cl (J. Bicking), 17cr (Matthew Gough), 20 (O Brasil que poucos conhecem), 21t (P Harstela), 23b (Eric Isselee), 25t (Peter Kirillov), 26–27b (PongMoj), 31 (Aleksandra Smirnova).

Library and Archives Canada Cataloguing in Publication

Title: Animal engineers / Izzi Howell.
Names: Howell, Izzi, author.
Description: Series statement: Astonishing animals |
Includes bibliographical references and index.
Identifiers: Canadiana (print) 20200155121 | Canadiana (ebook) 20200155123X |
ISBN 9780778769309 (hardcover) |
ISBN 9780778769361 (softcover) |
ISBN 9781427124364 (HTML)
Subjects: LCSH: Animals—Habitations—Juvenile literature. |
LCSH: Animal behavior—Juvenile
literature. | LCSH: Adaptation (Biology)—Juvenile literature.
Classification: LCC QL756 .H69 2020 | DDC j591.56/4—dc23

Library of Congress Cataloging-in-Publication Data

Names: Howell, Izzi, author.
Title: Animal engineers / Izzi Howell.
Description: New York : Crabtree Publishing Company, 2020. |
Series: Astonishing animals | Includes index.
Identifiers: LCCN 2019053193 (print) | LCCN 2019053194 (ebook) |
ISBN 9780778769309 (hardcover) |
ISBN 9780778769361 (paperback) |
ISBN 9781427124364 (ebook)
Subjects: LCSH: Animals--Habitations--Juvenile literature. |
Animal behavior--Juvenile literature. |
Adaptation (Biology)--Juvenile literature.
Classification: LCC QL756 .H87 2020 (print) | LCC QL756 (ebook) |
DDC 591.56/4--dc23
LC record available at https://lccn.loc.gov/2019053193
LC ebook record available at https://lccn.loc.gov/2019053194

Crabtree Publishing Company
www.crabtreebooks.com 1-800-387-7650

Printed in the U.S.A./022020/CG20200102

Published in Canada
Crabtree Publishing
616 Welland Ave.
St. Catharines, Ontario
L2M 5V6

Published in the United States
Crabtree Publishing
PMB 59051
350 Fifth Avenue, 59th Floor
New York, New York 10118

Published in the United Kingdom
Crabtree Publishing
Maritime House
Basin Road North, Hove
BN41 1WR

Published in Australia
Crabtree Publishing
Unit 3 – 5 Currumbin Court
Capalaba
QLD 4157

Table of contents

Animal engineers

Brilliant builders of the animal kingdom

When you think of animal engineers, what comes to mind? You might immediately think of birds and their nests, or beavers and their dams. But there are actually many other species that build surprising structures and change their habitats to meet their needs.

Special structures

Animals build structures for different reasons. Sometimes, it's to create a home or shelter for their young, such as the baya weaver bird (see pages 6–7). Other animals, such as spiders (see pages 16–17), build sneaky traps to catch **prey**. Some animals even build structures as a way of communicating with each other, such as the bowerbird (see pages 18–19).

This crafty hummingbird has built a nest as a shelter for its hungry chicks. The nest is made out of materials such as twigs and leaves.

Ecosystem engineers

As well as building shelters, animals change their environment in other ways, such as cutting down plants and changing the shape of the land. These species are called "ecosystem engineers." One animal, coral (see pages 28–29), even grows its own reef habitat!

North American beavers (see pages 20–21) are the lumberjacks of the animal kingdom, cutting down trees to build their dams. Beaver dams have a huge impact on rivers. They block the flow of water to create new lakes and to flood **wetland** areas.

Baya weaver birds

Unbelievable upside-down nests

If you turned a nest upside down, wouldn't the eggs fall out? Baya weaver birds build nests with an opening at the bottom, but their eggs stay safe because of the nest's clever design.

Wonderful weavers

Once a year, in **breeding** season, male baya weaver birds start to build nests for their **mates**. They cut long strips of grass and leaves using their sharp beaks. Then, they use their beaks to weave and knot the strips together. The birds build their nests from the top down.

Male and female baya weaver birds are differently colored. A female (above) is light brown, while males are yellow and brown.

FACT FILE

Found in: India and Southeast Asia

Habitat: Grasslands, farmed areas

Size: 6 inches (15 cm) long

Diet: Seeds and grains

Wow!

It can take a male baya weaver bird up to 500 trips to gather the materials for a nest!

The male baya weaver bird ties knots in the strips of grass and leaves to hold them in place.

The entrance to the nest is through a tunnel at the bottom. This makes it harder for predators to find their way into the nest.

Finding a mate

Once a nest is almost finished, the male baya weaver bird starts looking for a mate. He hangs off the nest and flaps his wings to attract the attention of female birds. Females are also attracted to the bright yellow feathers that male baya weaver birds grow during breeding season. Once a male has found a mate, he finishes the nest by building the tunnel. Then, the female lays eggs inside and raises the chicks.

Chimpanzees

New beds every night

Chimpanzee nests are no monkey business! Every night, chimpanzees build new nests in trees to sleep in.

Chimpanzee nests are 4–5 feet (1.2–1.5 m) long and around 3 feet (0.9 m) wide.

A chimpanzee can rustle up a new nest in around 10 minutes.

Chimpanzees line their nests with leaves and small branches to make them more comfortable. Sweet dreams!

Helping hands

Like humans, chimpanzees have **opposable thumbs**, which means they can do fiddly tasks with their hands. They weave branches together to create strong platforms for their nests, high in the treetops. Fussy chimps spend time searching for branches from specific trees, such as the Ugandan Ironwood, to build their nests. These trees have tough branches, so there's less chance of the nest breaking during the night!

FACT FILE

Found in: Africa

Habitat: Tropical forests

Size: 4–5.5 feet (1.2–1.7 m) tall

Diet: Fruit, leaves, eggs, insects, and occasional meat

During the day, chimpanzees scout out the perfect nest location.

They may reuse sites that have already been bent into the perfect nest shape!

Safe sleepers

Chimpanzees spend eight to nine hours a night in their nests. They are high off the ground, which keeps them safe from predators, such as lions, and away from irritating insects.

Wow!

There are lower levels of **bacteria** in chimpanzee nests than in human beds! Time to change your sheets?

Harvest mice

Tiny tennis ball-sized nests

What's the best house for a harvest mouse? These mini mice weave small, ball-shaped nests high among stalks of grass, out of the sight of predators.

Getting ready

Female harvest mice start work on their nest when they are pregnant. They tear strips of fresh grass using their front teeth and weave them into a ball around 3 inches (8 cm) in diameter. Each miniature nest is secured to a plant stalk, around 10–40 inches (25–102 cm) above the ground.

Fresh green grass is flexible and easier to weave. It also acts as camouflage for the nest!

No need to worry about the weight of the nest in the fragile grass—a harvest mouse only weighs about as much as five paperclips!

Preparing for a baby

It takes two or three days to finish the nest. Some females add a lining of grass or soft plant **fibers** to the nest to keep things cozy! They then give birth to their young inside.

FACT FILE

Found in: Europe and Asia

Habitat: Grasslands and farmland

Size: 2–3 inches (5–7.6 cm) long

Diet: Seeds, fruit, and insects

Harvest mice will happily nest inside a tennis ball, which is about the same size as a nest. Scientists set up these tennis-ball homes to help them track mouse populations.

Harvest mouse nests are so small that it's hard for predators to spot them among the plants.

Wow!

Harvest mice can hold on to plant stems and other objects with their tail, just like a monkey!

Wasps

Wonderful woodworkers

How do you build a nest from wood? By chewing it up and mixing it into a paste with saliva! Wasps use this lovely mixture to make strong nests for their young.

A new colony

Wasps can be split into two groups: social and **solitary**. Both groups build nests. In social wasp nests, the queen wasp starts the building process every spring when she wakes up from her **hibernation**. She builds a small nest and lays a few eggs in it to start a new colony.

Wasps build their nests anywhere—in trees, on walls, or even underground!

A wasp nest starts out small. Worker wasps add extra space in the nest as the colony gets bigger.

What's that sound? It's a wasp chewing through solid wood with its mighty mandibles, or jaws!

Hatching from honeycomb

Inside a social wasp nest are honeycomb-shaped cells made from chewed-up wood pulp. The queen wasp lays an egg in every cell. Worker wasps hatch from the first eggs laid by the queen. When they grow up, they help to maintain and expand the nest, while the queen lays more eggs.

Wow!

One wasp nest found in Tasmania in 2015 weighed 200 pounds (91 kg), which is as much as a large adult man!

Solitary wasps do not live in colonies. They build many types of nests. Some live in holes in trees, while others build their own nests from soil or chewed-up plant material.

Cathedral termites

Mega mounds

Have you ever visited a cathedral? These churches are often tall and elaborate. Cathedrals—natural ones, that is—are also home to tiny insects. Cathedral termites get their name from the huge, complex mounds that they build, which look a bit like cathedrals!

A mound of millions

Under every cathedral mound is a nest, which is home to millions of termites. The nest is damp, dark, and warm—just how termites like it! It also provides shelter from the hot Sun and protection from predators, such as ants, birds, and frogs.

Termite team

Termites are social animals. Every termite has its own role and tasks to carry out. Worker termites are responsible for building and maintaining the nest. Even though they are tiny, huge numbers of worker termites can build giant mounds by working as a team. Soldier termites defend the nest by attacking anything that breaks through the wall. They also spray the outside of the mound with stinky chemicals to put off predators.

FACT FILE

Found in: Australia

Habitat: Grasslands

Size: 0.1 inches (2.5 mm) tall for a worker termite

Diet: Grass

It's go time! Worker termites rush to mend any holes that appear in the mound.

At over 2,716 feet (828 m) tall, the Burj Khalifa is the tallest building on Earth!

Wow!

As termites are so small, the height of their nests is comparable to humans building a structure as tall as four Burj Khalifas stacked on top of each other!

Spiders

Super-strong silk traps

How can an insect become trapped by one thin strand of silk? It can if the strand is part of a spider web! Some spiders build intricate webs as traps to catch prey. Their webs may look fragile, but they are actually incredibly strong!

Wow!

A strand of spider silk is 1,000 times thinner than a human hair, but is five times stronger than steel!

Sticky or strong?

Spider webs are made from silk—a natural fiber that spiders produce from special silk **glands**. Each silk gland produces a different type of silk: sticky, strong, or thick. These silks are used to create different styles of web.

Different web designs

Orb web

Spiders lay the diagonal threads first, then add silk on top in a spiral shape. These hanging webs are used to catch flying insects.

Sheet web

This horizontal web lies on top of bushes or areas of grass. Strands of silk above the sheet web knock insects down into the web.

Tangle web

This random collection of threads is an abstract work of art! Tangle webs are often built under plants and dead leaves, so they catch crawling insects.

Funnel web

This web is formed of a sheet of silk, which comes out of a central tube. Funnel webs are built among grass and low plants to catch insects that live on the ground.

Spider attack

Some spiders make their webs from sticky silk that traps insects. The insects get tangled in the web as they struggle to escape. Other spiders use a special thread called the signal thread, which sends vibrations to the spider when an insect touches the web. The spider then darts out to attack its prey.

Spiders wrap captured insects in silk to keep them still! They then bite the insects to kill them and eat them.

Spider silk has many uses! It is also used to build nests for young spiders and sleeping chambers.

Bowerbirds

Colorful structures to charm females

Is that a work of art? No, it's a bowerbird bower! Male bowerbirds decorate incredible structures called bowers with petals, seeds, and stones to attract a mate.

Building the bower

As mating season approaches, male bowerbirds prepare their bowers. These are structures made from grass and sticks. It can take anywhere from a week to several months to prepare the perfect bower. Artistic male bowerbirds build different styles of bower, including walls of sticks or a small hut. Then, they use their beaks to add colorful accessories to attract the attention of the ladies!

Different species of bowerbird prefer different colors of decoration. However, most species don't match decoration color to their feather color.

Some bowerbirds are artists! They cover the walls of the bower with paint made from plant juices or charcoal dust mixed with saliva. They use their beak or a piece of bark as a paintbrush.

Bowers are decorated with flowers, fruit, seeds, small stones, dead insects, moss, and even garbage left behind by humans!

Attracting a mate

When a female bowerbird passes by the bower, the male entertains her with a song or a dance. She may visit several different bowers before choosing a male to mate with. Female bowerbirds lay their eggs in cup-shaped nests high in trees.

FACT FILE

Found in: New Guinea and Australia

Habitat: Tropical forests and woodlands

Size: 9–15 inches (23–38 cm) long

Diet: Fruit and insects

Some male bowerbirds maintain their bowers for 20 years, coming back to them to add new decorations every year.

The female bowerbird is often less brightly colored than the male.

Wow!

The bowers of the Vogelkop gardener bowerbird can be up to 39 inches (1 m) high!

North American beavers

Resourceful river dams

How can one animal change the path of an entire river? Beavers transform their surroundings by building dams in streams and rivers. The dams block the flow of water in the river, creating new, deep ponds.

Pool planners

Beaver dams block rivers and flood the area behind the dams. This creates deep pools that don't totally freeze during winter. This allows beavers to enter their lodges (see page 21) all year round. Predators can't enter these flooded wetland areas, which keeps beavers safe. The pools also become home to fish and amphibians.

A lovely lodge

A beaver's work isn't over once the dam is finished! Next, they build large shelters above the water called lodges. These are made from more of the beavers' favorite materials—branches and logs! The only entrance to a lodge is a tunnel that opens underwater.

FACT FILE

Found in: Canada, United States

Habitat: Rivers, marshes, and lakes

Size: 31 inches (79 cm) long

Diet: Tree bark, water plants

Lodges are well-**insulated** to keep beavers warm throughout the winter.

Inside the lodges, beavers are safe from predators such as bears.

Wow!

Beavers' front teeth are orange because they contain the metal iron. This metal makes the teeth strong enough to gnaw through trees!

African bush elephants

Helping by hurting

Do African bush elephants really help their habitat by destroying it? Unbelievably, the damage done by African bush elephants attracts other animals to the area!

Stomp, crash, thud!

African bush elephants are very large and strong, so they often do a lot of damage when they look for food. They pull up huge clumps of grass and dig huge holes in the ground looking for roots. This sounds like it would be bad for the habitat, but it actually creates space for new plants to grow!

Timber! Rather than trying to reach the delicious branches at the top, African bush elephants sometimes pull over trees at mealtime.

Adding animals

The wrecked landscape left by African bush elephants is the perfect place for animals, such as zebras and antelope, to hide from predators. Insects are also attracted to these areas, which in turn brings in animals that eat insects, such as amphibians and reptiles. This makes the habitat more **biodiverse**.

FACT FILE

Found in: Africa

Habitat: Grasslands

Size: 10–13 feet (3–4 m) tall

Diet: Tree branches and leaves, roots, and fruit

Elephants love to munch on trees, which means very few trees grow untouched on the grassland. With fewer trees, there is more space for grass and more food for grazing animals, such as zebras.

Elephants help transport plant seeds to new areas by eating them! The seeds pass through their system whole and are left behind in the elephant's dung, or poop.

Wow!

African bush elephants have huge appetites! One adult can eat over 285 pounds (129 kg) of food every day!

Black-tailed prairie dogs

Best of the burrowers

All dogs love digging holes. But did you know there's one that dives in and lives underground with its BFFs? It's the black-tailed prairie dog! These social animals live together in huge underground colonies.

Room for everyone!

Black-tailed prairie dogs build giant, underground colonies in a U-shape. Colonies have many different chambers connected by tunnels. Each chamber has a different purpose, such as a bedroom, a nursery, or an area to listen out for predators. Living underground protects prairie dogs from predators, rain, snow, and extreme temperatures.

Great grasslands

Black-tailed prairie dog colonies have a positive impact on grassland habitats. They create air pockets in the soil, which keeps it from drying out. Their burrows also provide shelter for other species, such as burrowing owls, who hide inside, away from predators or storms.

FACT FILE

Found in: North America

Habitat: Grasslands

Size: 11–13 inches (28–33 cm) long

Diet: Leaves, branches, seeds, and roots

Grazing prairie dogs help keep the grass short. This creates habitats for some bird species, such as the horned lark.

Lunchtime! Prairie dogs come out of the burrows to feed on grass and other plants.

Wow!

The largest recorded black-tailed prairie dog colony covered 25,000 square miles (64,750 square km)—slightly bigger than the state of West Virginia! It was home to 400 million prairie dogs!

Oysters

Brilliant habitat builders

Would you rather travel the seas or live your life in one place? Oysters are homebodies that spend almost their entire life fixed in one spot. The buildup of oysters creates an oyster reef!

Shells and structures

Oysters begin their lives as tiny creatures that can swim freely. After several days, they fix themselves to a solid surface and grow into their adult form. Oysters attach themselves to rocks, underwater structures, or old oyster shells. Over time, layers and layers of oysters build up, creating a structure known as an oyster reef.

This oyster reef is exposed at **low tide.**

Oysters and their reefs have been damaged by pollution, habitat loss, and over-harvesting for food. It takes decades for an oyster reef to recover from such damage.

Ocean assistance

Oyster reefs provide an important habitat for other species, and a good hiding place where fish and crabs can find shelter from predators. The reefs also act as a natural storm barrier, helping to break up large waves as they come close to land. Oysters themselves **filter** the water around them, helping to keep the ocean clean. The loss of oyster reefs is causing problems with biodiversity, water quality, and storm damage.

FACT FILE

Found in: Cool and warm oceans

Size: 2–14 inches (5–36 cm) long

Diet: Tiny pieces of natural matter and **plankton**

The Billion Oyster Project is working to create new oyster reefs in the waters around New York City. They make structures from old oyster shells or metal frames, hoping new oysters will attach themselves and form new reefs.

Many small animals, such as barnacles and marine snails, live on oyster shells or in the spaces between oysters in an oyster reef.

Wow!

An adult oyster filters up to 47.5 gallons (180 liters) of water every day!

Coral

Sensational sea structures

Did you know that coral reefs are made out of dead bodies? Coral bodies, that is! Over time, coral skeletons build up to form massive underwater structures.

This is a zoomed-in image of a coral polyp. A polyp is a small, tube-shaped animal.

This end of the coral polyp has tentacles to grab food.

The other end of the coral polyp is secured down to stop the polyp from floating away!

Plant or polyp?

Coral is often mistaken for a plant, but it is actually made up of many identical coral polyps. Each coral polyp has a hard outer skeleton made of **minerals**. When polyps die, more polyps grow on top of their empty skeletons. Over time, this process builds up a huge coral structure.

Some projects are planting young coral in damaged reefs to replace coral that has been destroyed.

Coral reefs are sometimes called "rain forests of the sea," because they are home to so many species, just like rain forests!

Coral in crisis

Coral reefs are under attack! They are threatened by many dangers, including pollution, damage from tourists, and high water temperatures caused by **global warming**. The destruction of coral reefs will result in the loss of many ocean animals and plants that live in reefs, such as seaweed, fish, and sea anemones.

Wow!

Coral reefs only cover 0.1% of the ocean, but they are home to 25% of all ocean species!

Danger! Too many crown of thorn starfish, pictured here, are invading reefs and eating coral.

Glossary

abdomen The lower part of an animal's body

bacteria Very small organisms, some of which can cause diseases

biodiverse Having many different species of plants and animals

breeding When animals reproduce and have young

colony A group of the same type of animals that live together

dam A structure built across a river that blocks the flow of water

fiber A structure or object that is shaped like a thread

filter To remove solids from a liquid or a gas

gland An organ in the body that produces substances, such as sweat, that have specific uses in the body

global warming The gradual increase in Earth's temperature caused by human activities, such as burning fossil fuels

grassland A large, grassy area of land

hibernation The state of being asleep or inactive for the winter

insulated Designed in a way so that heat does not escape

intricate Having many small and complicated parts and details

larvae A young, worm-like form of an insect

low tide When the tide, or rising and falling of the ocean's surface, has reached its lowest level and uncovered the land

mate An animal's reproductive partner

mineral A solid chemical substance found in nature

opposable thumbs Thumbs that allow you to pick up something between the thumb and first finger

plankton Very small plants and animals that live in the ocean

predator An animal that hunts another animal for food

prey An animal that is hunted and killed for food by other animals

saliva Spit

social Describes an animal that lives together with other animals of the same species in an organized way

solitary Describes an animal that lives alone, rather than with other animals of the same species

wetland A large area of soft, wet land

Find out more

Books

Gray, Leon. *Amazing Animal Engineers*. Capstone Press, 2015.

Scarborough, Kate and Camm, Martin. *A Wasp Builds a Nest*. Firefly Books, 2016.

Tornio, Stacy. *Bird Nests*. Focus Readers, 2018.

Websites

thekidshouldseethis.com/post/85620381170
Watch a video and see for yourself how North American beavers build their dams.

kids.nceas.ucsb.edu/biomes/coralreef.html
Read all about coral reefs and explore further with activities and helpful links.

www.wonderopolis.org/wonder/how-do-wasps-build-their-nests
Find out more about how wasps build their nests.

Index